THIS LOG BO(

Name: _____

Business Name: _____
_____

Address: _____
_____
_____

Website: _____

Email: _____

Cell Phone: _____

Phone: _____

Emergency contact: _____

Fax: _____

CONTENTS

COLONY SETUP/MAP .... PAGE 4-5
RECIPES .... PAGE 6-11
CARE OF BEES - NOTES/INSTRUCTIONS .... PAGE 13-14
SEASONAL TO DO LISTS .... PAGE 6-18
BEEKEEPING JOURNAL/LOG PAGES .... PAGE 19-103
BEEKEEPING INCOME .... PAGE 104-105
BEEKEEPING EXPENSES .... PAGE 106-107
BEEKEEPING EQUIPMENT .... PAGE 108
EQUIPMENT USE/CLEANING LOG .... PAGE 109
BLANK NOTES PAGE .... PAGE 110

COLONY SET UP
USE THIS SPACE TO MAP YOUR COLONY OR DRAW DETAILS OF YOUR COLONY SET UP

COLONY SET UP

USE THIS SPACE TO MAP YOUR COLONY OR DRAW DETAILS OF YOUR COLONY SET UP

BEE FOOD RECIPE

RECIPE SOURCE (FOR FUTURE REFERENCE): _____

PURPOSE (SEASONAL, REPAIR, ETC): _____

_____

INSTRUCTIONS FOR USE: _____

_____

AMT RECIPE MAKES: _____   PREP TIME: _____

INGREDIENTS: EQUIPMENT NEEDED:

_____   _____

_____   _____

_____   _____

_____   _____

_____   _____

_____   _____

INSTRUCTIONS:

_____

_____

_____

_____

_____

_____

_____

_____

_____

_____

_____

_____

_____

_____

NOTES:

_____

_____

_____

BEE FOOD RECIPE _____

RECIPE SOURCE (FOR FUTURE REFERENCE): _____

PURPOSE (SEASONAL, REPAIR, ETC): _____

INSTRUCTIONS FOR USE: _____

AMT RECIPE MAKES: _____   PREP TIME: _____

INGREDIENTS: EQUIPMENT NEEDED:

_____   _____

_____   _____

_____   _____

_____   _____

_____   _____

INSTRUCTIONS:

_____

_____

_____

_____

_____

_____

_____

_____

_____

_____

_____

_____

_____

_____

_____

NOTES:

_____

_____

BEE FOOD RECIPE

RECIPE SOURCE (FOR FUTURE REFERENCE): _____

PURPOSE (SEASONAL, REPAIR, ETC): _____

INSTRUCTIONS FOR USE: _____

AMT RECIPE MAKES: _____   PREP TIME: _____

INGREDIENTS: EQUIPMENT NEEDED:

INSTRUCTIONS:

NOTES:

BEE FOOD RECIPE _____

RECIPE SOURCE (FOR FUTURE REFERENCE): _____

PURPOSE (SEASONAL, REPAIR, ETC): _____

_____

INSTRUCTIONS FOR USE: _____

_____

AMT RECIPE MAKES: _____   PREP TIME: _____

INGREDIENTS: EQUIPMENT NEEDED:

_____   _____
_____   _____
_____   _____
_____   _____
_____   _____
_____   _____

INSTRUCTIONS:

_____
_____
_____
_____
_____
_____
_____
_____
_____
_____
_____
_____
_____
_____
_____

NOTES:

_____
_____
_____

BEE FOOD RECIPE

RECIPE SOURCE (FOR FUTURE REFERENCE): _____

PURPOSE (SEASONAL, REPAIR, ETC): _____

_____

INSTRUCTIONS FOR USE: _____

_____

AMT RECIPE MAKES: _____   PREP TIME: _____

INGREDIENTS: EQUIPMENT NEEDED:

_____   _____
_____   _____
_____   _____
_____   _____
_____   _____
_____

INSTRUCTIONS:

_____
_____
_____
_____
_____
_____
_____
_____
_____
_____
_____
_____
_____
_____
_____

NOTES:

_____
_____
_____

BEE FOOD RECIPE _____

RECIPE SOURCE (FOR FUTURE REFERENCE): _____

PURPOSE (SEASONAL, REPAIR, ETC): _____

_____

INSTRUCTIONS FOR USE: _____

_____

AMT RECIPE MAKES: _____   PREP TIME: _____

INGREDIENTS: EQUIPMENT NEEDED:

_____   _____

_____   _____

_____   _____

_____   _____

_____   _____

_____   _____

INSTRUCTIONS:

_____

_____

_____

_____

_____

_____

_____

_____

_____

_____

_____

_____

_____

NOTES:

_____

_____

_____

CARE OF BEES - NOTES AND INSTRUCTIONS

Note any routines and care instructions here. Useful for refining your processes or passing on to an employee, family member or bee babysitter, should you need to be away.

SEASONAL TO DO LIST - SPRING

SEASONAL TO DO LIST - SUMMER

SEASONAL TO DO LIST - FALL

SEASONAL TO DO LIST - WINTER

BEEKEEPING JOURNAL/LOG   HIVE ID:_____

DATE:_____ TIME:_____ WEATHER:_____

HIVE STRUCTURE: FRAMES:_____ SUPERS:_____
BROOD FRAMES:_____ HONEY FRAMES:_____ OPEN FRAMES:_____
HIVE CAPACITY NOTES:_____
_____

HIVE INSPECTION NOTES:
BEE MOOD/TEMPERAMENT:_____
POLLEN?_____ HONEY FLOW?_____
FOOD/WATER NOTES:_____
QUEEN? YES☐ NO☐   MARKED?:YES☐ NO☐   COLOR:_____
BROOD STAGE (EGG, LARVA, PUPA), PATTERN AND NOTES:_____
_____
_____

LARVAE STATUS AND NOTES: _____
_____
_____

COMB ABNORMALITIES (QUEEN CELLS/DRONE COMB BUILDING, ETC.):
_____
_____

SIGNS OF PESTS/PROBLEMS? (MITES, ANTS, MOTHS, DEAD BEES, SMELL):
_____
_____

TREATMENTS/MEDICATIONS: _____
_____
_____

HONEY STORES: _____
SEASONAL NOTES (BLOOM, POLLEN/NECTAR SOURCES):_____
_____
_____

RATING OF OVERALL HIVE HEALTH   ⬡   ⬡   ⬡   ⬡
                                 WEAK   AVERAGE   STRONG

NOTES/COMB DIAGRAM:

BEEKEEPING JOURNAL/LOG   HIVE ID:_____

DATE:_____ TIME:_____ WEATHER:_____

HIVE STRUCTURE: FRAMES:_____ SUPERS:_____
BROOD FRAMES:_____ HONEY FRAMES:_____ OPEN FRAMES:_____
HIVE CAPACITY NOTES:_____
_____

HIVE INSPECTION NOTES:
BEE MOOD/TEMPERAMENT:_____
POLLEN?_____ HONEY FLOW?_____
FOOD/WATER NOTES: _____
QUEEN? YES ☐ NO ☐   MARKED?: YES ☐ NO ☐   COLOR:_____
BROOD STAGE (EGG, LARVA, PUPA), PATTERN AND NOTES:_____
_____
_____

LARVAE STATUS AND NOTES: _____
_____
_____

COMB ABNORMALITIES (QUEEN CELLS/DRONE COMB BUILDING, ETC.):
_____
_____

SIGNS OF PESTS/PROBLEMS? (MITES, ANTS, MOTHS, DEAD BEES, SMELL):
_____
_____

TREATMENTS/MEDICATIONS: _____
_____
_____

HONEY STORES: _____
SEASONAL NOTES (BLOOM, POLLEN/NECTAR SOURCES):_____
_____

RATING OF OVERALL HIVE HEALTH   ⬡ ⬡ ⬡ ⬡ ⬡
                                  WEAK   AVERAGE   STRONG

NOTES/COMB DIAGRAM:

BEEKEEPING JOURNAL/LOG HIVE ID: _____

DATE: _____ TIME: _____ WEATHER: _____

HIVE STRUCTURE: FRAMES: _____ SUPERS: _____
BROOD FRAMES: _____ HONEY FRAMES: _____ OPEN FRAMES: _____
HIVE CAPACITY NOTES: _____
_____

HIVE INSPECTION NOTES:
BEE MOOD/TEMPERAMENT: _____
POLLEN? _____ HONEY FLOW? _____
FOOD/WATER NOTES: _____
QUEEN? YES ☐ NO ☐   MARKED?: YES ☐ NO ☐   COLOR: _____
BROOD STAGE (EGG, LARVA, PUPA), PATTERN AND NOTES: _____
_____
_____

LARVAE STATUS AND NOTES: _____
_____
_____

COMB ABNORMALITIES (QUEEN CELLS/DRONE COMB BUILDING, ETC.):
_____
_____

SIGNS OF PESTS/PROBLEMS? (MITES, ANTS, MOTHS, DEAD BEES, SMELL):
_____
_____

TREATMENTS/MEDICATIONS: _____
_____

HONEY STORES: _____
SEASONAL NOTES (BLOOM, POLLEN/NECTAR SOURCES): _____
_____

RATING OF OVERALL HIVE HEALTH   ⬡   ⬡   ⬡   ⬡
                                  WEAK   AVERAGE   STRONG

23

NOTES/COMB DIAGRAM:

BEEKEEPING JOURNAL/LOG HIVE ID: _____

DATE: _____ TIME: _____ WEATHER: _____

HIVE STRUCTURE: FRAMES: _____ SUPERS: _____

BROOD FRAMES: _____ HONEY FRAMES: _____ OPEN FRAMES: _____

HIVE CAPACITY NOTES: _____
_____

HIVE INSPECTION NOTES:

BEE MOOD/TEMPERAMENT: _____
POLLEN? _____ HONEY FLOW? _____
FOOD/WATER NOTES: _____
QUEEN? YES ☐ NO ☐   MARKED?: YES ☐ NO ☐   COLOR: _____
BROOD STAGE (EGG, LARVA, PUPA), PATTERN AND NOTES: _____
_____
_____

LARVAE STATUS AND NOTES: _____
_____
_____

COMB ABNORMALITIES (QUEEN CELLS/DRONE COMB BUILDING, ETC.):
_____
_____

SIGNS OF PESTS/PROBLEMS? (MITES, ANTS, MOTHS, DEAD BEES, SMELL):
_____
_____

TREATMENTS/MEDICATIONS: _____
_____
_____

HONEY STORES: _____
SEASONAL NOTES (BLOOM, POLLEN/NECTAR SOURCES): _____
_____
_____

RATING OF OVERALL HIVE HEALTH ⬡ ⬡ ⬡ ⬡ ⬡
                                  WEAK   AVERAGE   STRONG

NOTES/COMB DIAGRAM:

BEEKEEPING JOURNAL/LOG   HIVE ID: _____

DATE: _____   TIME: _____   WEATHER: _____

HIVE STRUCTURE: FRAMES: _____ SUPERS: _____

BROOD FRAMES: _____ HONEY FRAMES: _____ OPEN FRAMES: _____

HIVE CAPACITY NOTES: _____
_____

HIVE INSPECTION NOTES:

BEE MOOD/TEMPERAMENT: _____
POLLEN?_____ HONEY FLOW? _____
FOOD/WATER NOTES: _____
QUEEN? YES☐ NO☐   MARKED?:YES☐ NO☐   COLOR: _____
BROOD STAGE (EGG, LARVA, PUPA), PATTERN AND NOTES: _____
_____
_____

LARVAE STATUS AND NOTES: _____
_____
_____

COMB ABNORMALITIES (QUEEN CELLS/DRONE COMB BUILDING, ETC.):
_____
_____

SIGNS OF PESTS/PROBLEMS? (MITES, ANTS, MOTHS, DEAD BEES, SMELL):
_____
_____

TREATMENTS/MEDICATIONS: _____
_____
_____

HONEY STORES: _____
SEASONAL NOTES (BLOOM, POLLEN/NECTAR SOURCES):_____
_____
_____

RATING OF OVERALL HIVE HEALTH   ⬡ ⬡ ⬡ ⬡ ⬡
                                  WEAK   AVERAGE   STRONG

NOTES/COMB DIAGRAM:

BEEKEEPING JOURNAL/LOG HIVE ID:_____

DATE: _____ TIME: _____ WEATHER: _____

HIVE STRUCTURE: FRAMES:_____ SUPERS: _____

BROOD FRAMES: _____ HONEY FRAMES: _____ OPEN FRAMES: _____

HIVE CAPACITY NOTES:_____
_____

HIVE INSPECTION NOTES:

BEE MOOD/TEMPERAMENT:_____
POLLEN?_____ HONEY FLOW?_____
FOOD/WATER NOTES: _____
QUEEN? YES ☐ NO ☐   MARKED?: YES ☐ NO ☐   COLOR:_____
BROOD STAGE (EGG, LARVA, PUPA), PATTERN AND NOTES:_____
_____
_____

LARVAE STATUS AND NOTES: _____
_____
_____

COMB ABNORMALITIES (QUEEN CELLS/DRONE COMB BUILDING, ETC.):
_____
_____

SIGNS OF PESTS/PROBLEMS? (MITES, ANTS, MOTHS, DEAD BEES, SMELL):
_____
_____

TREATMENTS/MEDICATIONS: _____
_____
_____

HONEY STORES: _____
SEASONAL NOTES (BLOOM, POLLEN/NECTAR SOURCES):_____
_____

RATING OF OVERALL HIVE HEALTH ⬡ ⬡ ⬡ ⬡ ⬡
                                      WEAK   AVERAGE   STRONG

NOTES/COMB DIAGRAM:

BEEKEEPING JOURNAL/LOG HIVE ID:_____

DATE:_____ TIME:_____ WEATHER:_____

HIVE STRUCTURE: FRAMES:_____ SUPERS:_____
BROOD FRAMES:_____ HONEY FRAMES:_____ OPEN FRAMES:_____
HIVE CAPACITY NOTES:_____
_____

HIVE INSPECTION NOTES:
BEE MOOD/TEMPERAMENT:_____
POLLEN?_____ HONEY FLOW?_____
FOOD/WATER NOTES:_____
QUEEN? YES ☐ NO ☐   MARKED?: YES ☐ NO ☐   COLOR:_____
BROOD STAGE (EGG, LARVA, PUPA), PATTERN AND NOTES:_____
_____
_____

LARVAE STATUS AND NOTES:_____
_____
_____

COMB ABNORMALITIES (QUEEN CELLS/DRONE COMB BUILDING, ETC.):
_____
_____

SIGNS OF PESTS/PROBLEMS? (MITES, ANTS, MOTHS, DEAD BEES, SMELL):
_____
_____

TREATMENTS/MEDICATIONS:_____
_____
_____

HONEY STORES:_____
SEASONAL NOTES (BLOOM, POLLEN/NECTAR SOURCES):_____
_____
_____

RATING OF OVERALL HIVE HEALTH ⬡ ⬡ ⬡ ⬡ ⬡
                              WEAK   AVERAGE   STRONG

NOTES/COMB DIAGRAM:

BEEKEEPING JOURNAL/LOG HIVE ID: _____

DATE: _____ TIME: _____ WEATHER: _____

HIVE STRUCTURE: FRAMES: _____ SUPERS: _____

BROOD FRAMES: _____ HONEY FRAMES: _____ OPEN FRAMES: _____

HIVE CAPACITY NOTES: _____
_____

HIVE INSPECTION NOTES:

BEE MOOD/TEMPERAMENT: _____
POLLEN? _____ HONEY FLOW? _____
FOOD/WATER NOTES: _____
QUEEN? YES ☐ NO ☐   MARKED?: YES ☐ NO ☐   COLOR: _____
BROOD STAGE (EGG, LARVA, PUPA), PATTERN AND NOTES: _____
_____
_____

LARVAE STATUS AND NOTES: _____
_____
_____

COMB ABNORMALITIES (QUEEN CELLS/DRONE COMB BUILDING, ETC.):
_____
_____

SIGNS OF PESTS/PROBLEMS? (MITES, ANTS, MOTHS, DEAD BEES, SMELL):
_____
_____

TREATMENTS/MEDICATIONS: _____
_____
_____

HONEY STORES: _____
SEASONAL NOTES (BLOOM, POLLEN/NECTAR SOURCES): _____
_____
_____

RATING OF OVERALL HIVE HEALTH ⬡ ⬡ ⬡ ⬡
                                    WEAK   AVERAGE   STRONG

NOTES/COMB DIAGRAM:

BEEKEEPING JOURNAL/LOG   HIVE ID: _____

DATE: _____ TIME: _____ WEATHER: _____

HIVE STRUCTURE: FRAMES: _____ SUPERS: _____

BROOD FRAMES: _____ HONEY FRAMES: _____ OPEN FRAMES: _____

HIVE CAPACITY NOTES: _____
_____

HIVE INSPECTION NOTES:

BEE MOOD/TEMPERAMENT: _____
POLLEN? _____ HONEY FLOW? _____
FOOD/WATER NOTES: _____
QUEEN? YES ☐ NO ☐   MARKED?: YES ☐ NO ☐   COLOR: _____
BROOD STAGE (EGG, LARVA, PUPA), PATTERN AND NOTES: _____
_____
_____

LARVAE STATUS AND NOTES: _____
_____
_____

COMB ABNORMALITIES (QUEEN CELLS/DRONE COMB BUILDING, ETC.):
_____
_____

SIGNS OF PESTS/PROBLEMS? (MITES, ANTS, MOTHS, DEAD BEES, SMELL):
_____
_____

TREATMENTS/MEDICATIONS: _____
_____
_____

HONEY STORES: _____
SEASONAL NOTES (BLOOM, POLLEN/NECTAR SOURCES): _____
_____
_____

RATING OF OVERALL HIVE HEALTH   ⬡ ⬡ ⬡ ⬡ ⬡
                                  WEAK   AVERAGE   STRONG

NOTES/COMB DIAGRAM:

BEEKEEPING JOURNAL/LOG   HIVE ID: _____

DATE: _____   TIME: _____   WEATHER: _____

HIVE STRUCTURE: FRAMES: _____   SUPERS: _____

BROOD FRAMES: _____   HONEY FRAMES: _____   OPEN FRAMES: _____

HIVE CAPACITY NOTES: _____
_____

HIVE INSPECTION NOTES:

BEE MOOD/TEMPERAMENT: _____
POLLEN? _____   HONEY FLOW? _____
FOOD/WATER NOTES: _____
QUEEN? YES ☐ NO ☐   MARKED?: YES ☐ NO ☐   COLOR: _____
BROOD STAGE (EGG, LARVA, PUPA), PATTERN AND NOTES: _____
_____
_____

LARVAE STATUS AND NOTES: _____
_____
_____

COMB ABNORMALITIES (QUEEN CELLS/DRONE COMB BUILDING, ETC.):
_____
_____

SIGNS OF PESTS/PROBLEMS? (MITES, ANTS, MOTHS, DEAD BEES, SMELL):
_____
_____

TREATMENTS/MEDICATIONS: _____
_____

HONEY STORES: _____
SEASONAL NOTES (BLOOM, POLLEN/NECTAR SOURCES): _____
_____

RATING OF OVERALL HIVE HEALTH   ⬡ ⬡ ⬡ ⬡ ⬡
                                  WEAK   AVERAGE   STRONG

NOTES/COMB DIAGRAM:

BEEKEEPING JOURNAL/LOG   HIVE ID: _____

DATE: _____   TIME: _____   WEATHER: _____

HIVE STRUCTURE: FRAMES: _____   SUPERS: _____

BROOD FRAMES: _____   HONEY FRAMES: _____   OPEN FRAMES: _____

HIVE CAPACITY NOTES: _____
_____

HIVE INSPECTION NOTES:

BEE MOOD/TEMPERAMENT: _____
POLLEN? _____   HONEY FLOW? _____
FOOD/WATER NOTES: _____
QUEEN? YES ☐ NO ☐   MARKED?: YES ☐ NO ☐   COLOR: _____
BROOD STAGE (EGG, LARVA, PUPA), PATTERN AND NOTES: _____
_____
_____

LARVAE STATUS AND NOTES: _____
_____
_____

COMB ABNORMALITIES (QUEEN CELLS/DRONE COMB BUILDING, ETC.):
_____
_____

SIGNS OF PESTS/PROBLEMS? (MITES, ANTS, MOTHS, DEAD BEES, SMELL):
_____
_____

TREATMENTS/MEDICATIONS: _____
_____
_____

HONEY STORES: _____
SEASONAL NOTES (BLOOM, POLLEN/NECTAR SOURCES): _____
_____
_____

RATING OF OVERALL HIVE HEALTH   ⬡ ⬡ ⬡ ⬡ ⬡
                                  WEAK   AVERAGE   STRONG

NOTES/COMB DIAGRAM:

BEEKEEPING JOURNAL/LOG   HIVE ID: _____

DATE: _____   TIME: _____   WEATHER: _____

HIVE STRUCTURE: FRAMES: _____   SUPERS: _____

BROOD FRAMES: _____   HONEY FRAMES: _____   OPEN FRAMES: _____

HIVE CAPACITY NOTES: _____
_____

HIVE INSPECTION NOTES:

BEE MOOD/TEMPERAMENT: _____
POLLEN? _____   HONEY FLOW? _____
FOOD/WATER NOTES: _____
QUEEN? YES ☐ NO ☐   MARKED?: YES ☐ NO ☐   COLOR: _____
BROOD STAGE (EGG, LARVA, PUPA), PATTERN AND NOTES: _____
_____
_____

LARVAE STATUS AND NOTES: _____
_____
_____

COMB ABNORMALITIES (QUEEN CELLS/DRONE COMB BUILDING, ETC.):
_____
_____

SIGNS OF PESTS/PROBLEMS? (MITES, ANTS, MOTHS, DEAD BEES, SMELL):
_____
_____

TREATMENTS/MEDICATIONS: _____
_____
_____

HONEY STORES: _____
SEASONAL NOTES (BLOOM, POLLEN/NECTAR SOURCES): _____
_____

RATING OF OVERALL HIVE HEALTH   ⬡ ⬡ ⬡ ⬡ ⬡
                                    WEAK   AVERAGE   STRONG

NOTES/COMB DIAGRAM:

BEEKEEPING JOURNAL/LOG   HIVE ID: _____

DATE: _____   TIME: _____   WEATHER: _____

HIVE STRUCTURE: FRAMES: _____   SUPERS: _____

BROOD FRAMES: _____   HONEY FRAMES: _____   OPEN FRAMES: _____

HIVE CAPACITY NOTES: _____
_____

HIVE INSPECTION NOTES:

BEE MOOD/TEMPERAMENT: _____
POLLEN? _____   HONEY FLOW? _____
FOOD/WATER NOTES: _____
QUEEN? YES ☐ NO ☐   MARKED?: YES ☐ NO ☐   COLOR: _____
BROOD STAGE (EGG, LARVA, PUPA), PATTERN AND NOTES: _____
_____
_____

LARVAE STATUS AND NOTES: _____
_____
_____

COMB ABNORMALITIES (QUEEN CELLS/DRONE COMB BUILDING, ETC.): _____
_____
_____

SIGNS OF PESTS/PROBLEMS? (MITES, ANTS, MOTHS, DEAD BEES, SMELL): _
_____
_____

TREATMENTS/MEDICATIONS: _____
_____

HONEY STORES: _____
SEASONAL NOTES (BLOOM, POLLEN/NECTAR SOURCES): _____
_____

RATING OF OVERALL HIVE HEALTH   ⬡ ⬡ ⬡ ⬡ ⬡
                                  WEAK   AVERAGE   STRONG

NOTES/COMB DIAGRAM:

BEEKEEPING JOURNAL/LOG HIVE ID:_____

DATE: _____ TIME: _____ WEATHER: _____

HIVE STRUCTURE: FRAMES:_____ SUPERS: _____

BROOD FRAMES: _____ HONEY FRAMES: _____ OPEN FRAMES: _____

HIVE CAPACITY NOTES:_____
_____

HIVE INSPECTION NOTES:

BEE MOOD/TEMPERAMENT: _____
POLLEN?_____ HONEY FLOW? _____
FOOD/WATER NOTES: _____
QUEEN? YES☐ NO☐   MARKED?:YES☐ NO☐   COLOR: _____
BROOD STAGE (EGG, LARVA, PUPA), PATTERN AND NOTES: _____
_____
_____

LARVAE STATUS AND NOTES: _____
_____
_____

COMB ABNORMALITIES (QUEEN CELLS/DRONE COMB BUILDING, ETC.):
_____
_____

SIGNS OF PESTS/PROBLEMS? (MITES, ANTS, MOTHS, DEAD BEES, SMELL):
_____
_____

TREATMENTS/MEDICATIONS: _____
_____
_____

HONEY STORES: _____
SEASONAL NOTES (BLOOM, POLLEN/NECTAR SOURCES):_____
_____
_____

RATING OF OVERALL HIVE HEALTH ⬡ ⬡ ⬡ ⬡ ⬡
                                    WEAK   AVERAGE   STRONG

NOTES/COMB DIAGRAM:

BEEKEEPING JOURNAL/LOG   HIVE ID: _____

DATE: _____   TIME: _____   WEATHER: _____

HIVE STRUCTURE: FRAMES: _____   SUPERS: _____

BROOD FRAMES: _____   HONEY FRAMES: _____   OPEN FRAMES: _____

HIVE CAPACITY NOTES: _____
_____

HIVE INSPECTION NOTES:

BEE MOOD/TEMPERAMENT: _____
POLLEN? _____   HONEY FLOW? _____
FOOD/WATER NOTES: _____
QUEEN? YES ☐ NO ☐   MARKED?: YES ☐ NO ☐   COLOR: _____
BROOD STAGE (EGG, LARVA, PUPA), PATTERN AND NOTES: _____
_____
_____

LARVAE STATUS AND NOTES: _____
_____
_____

COMB ABNORMALITIES (QUEEN CELLS/DRONE COMB BUILDING, ETC.):
_____
_____

SIGNS OF PESTS/PROBLEMS? (MITES, ANTS, MOTHS, DEAD BEES, SMELL):
_____
_____

TREATMENTS/MEDICATIONS: _____
_____
_____

HONEY STORES: _____
SEASONAL NOTES (BLOOM, POLLEN/NECTAR SOURCES): _____
_____

RATING OF OVERALL HIVE HEALTH   ⬡ ⬡ ⬡ ⬡ ⬡
                                   WEAK   AVERAGE   STRONG

NOTES/COMB DIAGRAM:

BEEKEEPING JOURNAL/LOG HIVE ID: _____

DATE: _____ TIME: _____ WEATHER: _____

HIVE STRUCTURE: FRAMES: _____ SUPERS: _____

BROOD FRAMES: _____ HONEY FRAMES: _____ OPEN FRAMES: _____

HIVE CAPACITY NOTES: _____
_____

HIVE INSPECTION NOTES:

BEE MOOD/TEMPERAMENT: _____
POLLEN? _____ HONEY FLOW? _____
FOOD/WATER NOTES: _____
QUEEN? YES ☐ NO ☐   MARKED? YES ☐ NO ☐   COLOR: _____
BROOD STAGE (EGG, LARVA, PUPA), PATTERN AND NOTES: _____
_____
_____

LARVAE STATUS AND NOTES: _____
_____
_____

COMB ABNORMALITIES (QUEEN CELLS/DRONE COMB BUILDING, ETC.):
_____
_____

SIGNS OF PESTS/PROBLEMS? (MITES, ANTS, MOTHS, DEAD BEES, SMELL):
_____
_____

TREATMENTS/MEDICATIONS: _____
_____
_____

HONEY STORES: _____
SEASONAL NOTES (BLOOM, POLLEN/NECTAR SOURCES): _____
_____
_____

RATING OF OVERALL HIVE HEALTH   ⬡   ⬡   ⬡   ⬡   ⬡
                                   WEAK   AVERAGE   STRONG

NOTES/COMB DIAGRAM:

BEEKEEPING JOURNAL/LOG   HIVE ID:_____

DATE: _____   TIME: _____   WEATHER: _____

HIVE STRUCTURE: FRAMES:_____ SUPERS: _____
BROOD FRAMES: _____ HONEY FRAMES: _____ OPEN FRAMES: _____
HIVE CAPACITY NOTES:_____
_____

HIVE INSPECTION NOTES:
BEE MOOD/TEMPERAMENT:_____
POLLEN?_____ HONEY FLOW? _____
FOOD/WATER NOTES: _____
QUEEN? YES☐ NO☐   MARKED?:YES☐ NO☐   COLOR:_____
BROOD STAGE (EGG, LARVA, PUPA), PATTERN AND NOTES: _____
_____
_____

LARVAE STATUS AND NOTES: _____
_____
_____

COMB ABNORMALITIES (QUEEN CELLS/DRONE COMB BUILDING, ETC.):
_____
_____

SIGNS OF PESTS/PROBLEMS? (MITES, ANTS, MOTHS, DEAD BEES, SMELL):
_____
_____

TREATMENTS/MEDICATIONS: _____
_____

HONEY STORES: _____
SEASONAL NOTES (BLOOM, POLLEN/NECTAR SOURCES):_____
_____

RATING OF OVERALL HIVE HEALTH   ⬡ ⬡ ⬡ ⬡ ⬡
                                        WEAK   AVERAGE   STRONG

NOTES/COMB DIAGRAM:

BEEKEEPING JOURNAL/LOG   HIVE ID: _____

DATE: _____ TIME: _____ WEATHER: _____

HIVE STRUCTURE: FRAMES: _____ SUPERS: _____

BROOD FRAMES: _____ HONEY FRAMES: _____ OPEN FRAMES: _____

HIVE CAPACITY NOTES: _____
_____

HIVE INSPECTION NOTES:

BEE MOOD/TEMPERAMENT: _____
POLLEN? _____ HONEY FLOW? _____
FOOD/WATER NOTES: _____
QUEEN? YES ☐ NO ☐   MARKED?: YES ☐ NO ☐   COLOR: _____
BROOD STAGE (EGG, LARVA, PUPA), PATTERN AND NOTES: _____
_____
_____

LARVAE STATUS AND NOTES: _____
_____
_____

COMB ABNORMALITIES (QUEEN CELLS/DRONE COMB BUILDING, ETC.):
_____
_____

SIGNS OF PESTS/PROBLEMS? (MITES, ANTS, MOTHS, DEAD BEES, SMELL):
_____
_____

TREATMENTS/MEDICATIONS: _____
_____
_____

HONEY STORES: _____
SEASONAL NOTES (BLOOM, POLLEN/NECTAR SOURCES): _____
_____
_____

RATING OF OVERALL HIVE HEALTH   ⬡ ⬡ ⬡ ⬡ ⬡
WEAK   AVERAGE   STRONG

NOTES/COMB DIAGRAM:

BEEKEEPING JOURNAL/LOG   HIVE ID: _____

DATE: _____   TIME: _____   WEATHER: _____

HIVE STRUCTURE: FRAMES: _____   SUPERS: _____

BROOD FRAMES: _____   HONEY FRAMES: _____   OPEN FRAMES: _____

HIVE CAPACITY NOTES: _____
_____

HIVE INSPECTION NOTES:

BEE MOOD/TEMPERAMENT: _____
POLLEN? _____   HONEY FLOW? _____
FOOD/WATER NOTES: _____
QUEEN? YES ☐ NO ☐   MARKED?: YES ☐ NO ☐   COLOR: _____
BROOD STAGE (EGG, LARVA, PUPA), PATTERN AND NOTES: _____
_____
_____

LARVAE STATUS AND NOTES: _____
_____

COMB ABNORMALITIES (QUEEN CELLS/DRONE COMB BUILDING, ETC.):
_____
_____

SIGNS OF PESTS/PROBLEMS? (MITES, ANTS, MOTHS, DEAD BEES, SMELL):
_____
_____

TREATMENTS/MEDICATIONS: _____
_____

HONEY STORES: _____
SEASONAL NOTES (BLOOM, POLLEN/NECTAR SOURCES): _____
_____

RATING OF OVERALL HIVE HEALTH   ⬡   ⬡   ⬡   ⬡
                                  WEAK   AVERAGE   STRONG

NOTES/COMB DIAGRAM:

BEEKEEPING JOURNAL/LOG HIVE ID: _____

DATE: _____ TIME: _____ WEATHER: _____

HIVE STRUCTURE: FRAMES: _____ SUPERS: _____

BROOD FRAMES: _____ HONEY FRAMES: _____ OPEN FRAMES: _____

HIVE CAPACITY NOTES: _____
_____

HIVE INSPECTION NOTES:

BEE MOOD/TEMPERAMENT: _____
POLLEN: _____ HONEY FLOW? _____
FOOD/WATER NOTES: _____
QUEEN? YES ☐ NO ☐   MARKED?: YES ☐ NO ☐   COLOR: _____
BROOD STAGE (EGG, LARVA, PUPA), PATTERN AND NOTES: _____
_____
_____

LARVAE STATUS AND NOTES: _____
_____
_____

COMB ABNORMALITIES (QUEEN CELLS/DRONE COMB BUILDING, ETC.):
_____
_____

SIGNS OF PESTS/PROBLEMS? (MITES, ANTS, MOTHS, DEAD BEES, SMELL):
_____
_____

TREATMENTS/MEDICATIONS: _____
_____
_____

HONEY STORES: _____

SEASONAL NOTES (BLOOM, POLLEN/NECTAR SOURCES): _____
_____
_____

RATING OF OVERALL HIVE HEALTH ⬡ ⬡ ⬡ ⬡ ⬡
                                  WEAK   AVERAGE   STRONG

NOTES/COMB DIAGRAM:

BEEKEEPING JOURNAL/LOG   HIVE ID: _____

DATE: _____   TIME: _____   WEATHER: _____

HIVE STRUCTURE: FRAMES: _____   SUPERS: _____

BROOD FRAMES: _____   HONEY FRAMES: _____   OPEN FRAMES: _____

HIVE CAPACITY NOTES: _____

_____

HIVE INSPECTION NOTES:

BEE MOOD/TEMPERAMENT: _____

POLLEN? _____   HONEY FLOW? _____

FOOD/WATER NOTES: _____

QUEEN? YES ☐ NO ☐   MARKED?: YES ☐ NO ☐   COLOR: _____

BROOD STAGE (EGG, LARVA, PUPA), PATTERN AND NOTES: _____

_____

_____

LARVAE STATUS AND NOTES: _____

_____

_____

COMB ABNORMALITIES (QUEEN CELLS/DRONE COMB BUILDING, ETC.):

_____

_____

SIGNS OF PESTS/PROBLEMS? (MITES, ANTS, MOTHS, DEAD BEES, SMELL):

_____

_____

TREATMENTS/MEDICATIONS: _____

_____

_____

HONEY STORES: _____

SEASONAL NOTES (BLOOM, POLLEN/NECTAR SOURCES): _____

_____

_____

RATING OF OVERALL HIVE HEALTH   ⬡   ⬡   ⬡   ⬡
                                  WEAK   AVERAGE   STRONG

NOTES/COMB DIAGRAM:

BEEKEEPING JOURNAL/LOG   HIVE ID:_____

DATE: _____   TIME: _____   WEATHER: _____

HIVE STRUCTURE: FRAMES:_____   SUPERS: _____

BROOD FRAMES: _____   HONEY FRAMES: _____   OPEN FRAMES: _____

HIVE CAPACITY NOTES:_____
_____

HIVE INSPECTION NOTES:

BEE MOOD/TEMPERAMENT:_____
POLLEN?_____   HONEY FLOW? _____
FOOD/WATER NOTES: _____
QUEEN? YES ☐ NO ☐   MARKED?: YES ☐ NO ☐   COLOR:_____
BROOD STAGE (EGG, LARVA, PUPA), PATTERN AND NOTES: _____
_____
_____

LARVAE STATUS AND NOTES: _____
_____
_____

COMB ABNORMALITIES (QUEEN CELLS/DRONE COMB BUILDING, ETC.):
_____
_____

SIGNS OF PESTS/PROBLEMS? (MITES, ANTS, MOTHS, DEAD BEES, SMELL):
_____
_____

TREATMENTS/MEDICATIONS: _____
_____
_____

HONEY STORES: _____
SEASONAL NOTES (BLOOM, POLLEN/NECTAR SOURCES):_____
_____

RATING OF OVERALL HIVE HEALTH   ◯ ◯ ◯ ◯ ◯
                                   WEAK   AVERAGE   STRONG

NOTES/COMB DIAGRAM:

BEEKEEPING JOURNAL/LOG HIVE ID:_____

DATE: _____ TIME: _____ WEATHER: _____

HIVE STRUCTURE: FRAMES:_____ SUPERS: _____

BROOD FRAMES: _____ HONEY FRAMES: _____ OPEN FRAMES: _____

HIVE CAPACITY NOTES:_____
_____

HIVE INSPECTION NOTES:

BEE MOOD/TEMPERAMENT:_____
POLLEN?_____ HONEY FLOW? _____
FOOD/WATER NOTES: _____
QUEEN? YES ☐ NO ☐   MARKED?:YES ☐ NO ☐   COLOR:_____
BROOD STAGE (EGG, LARVA, PUPA), PATTERN AND NOTES: _____
_____
_____

LARVAE STATUS AND NOTES: _____
_____
_____

COMB ABNORMALITIES (QUEEN CELLS/DRONE COMB BUILDING, ETC.):
_____
_____

SIGNS OF PESTS/PROBLEMS? (MITES, ANTS, MOTHS, DEAD BEES, SMELL):
_____
_____

TREATMENTS/MEDICATIONS: _____
_____

HONEY STORES: _____
SEASONAL NOTES (BLOOM, POLLEN/NECTAR SOURCES):_____
_____
_____

RATING OF OVERALL HIVE HEALTH   ◯ ◯ ◯ ◯ ◯
                                    WEAK   AVERAGE   STRONG

NOTES/COMB DIAGRAM:

BEEKEEPING JOURNAL/LOG   HIVE ID: _____

DATE: _____   TIME: _____   WEATHER: _____

HIVE STRUCTURE: FRAMES: _____   SUPERS: _____

BROOD FRAMES: _____   HONEY FRAMES: _____   OPEN FRAMES: _____

HIVE CAPACITY NOTES: _____
_____

HIVE INSPECTION NOTES:

BEE MOOD/TEMPERAMENT: _____
POLLEN? _____   HONEY FLOW? _____
FOOD/WATER NOTES: _____
QUEEN? YES ☐ NO ☐   MARKED?: YES ☐ NO ☐   COLOR: _____
BROOD STAGE (EGG, LARVA, PUPA), PATTERN AND NOTES: _____
_____
_____

LARVAE STATUS AND NOTES: _____
_____
_____

COMB ABNORMALITIES (QUEEN CELLS/DRONE COMB BUILDING, ETC.):
_____
_____

SIGNS OF PESTS/PROBLEMS? (MITES, ANTS, MOTHS, DEAD BEES, SMELL):
_____
_____

TREATMENTS/MEDICATIONS: _____
_____
_____

HONEY STORES: _____
SEASONAL NOTES (BLOOM, POLLEN/NECTAR SOURCES): _____
_____

RATING OF OVERALL HIVE HEALTH   ⬡   ⬡   ⬡   ⬡   ⬡
                                      WEAK   AVERAGE   STRONG

NOTES/COMB DIAGRAM:

BEEKEEPING JOURNAL/LOG   HIVE ID:_____

DATE: _____   TIME: _____   WEATHER: _____

HIVE STRUCTURE: FRAMES:_____   SUPERS: _____

BROOD FRAMES: _____   HONEY FRAMES: _____   OPEN FRAMES: _____

HIVE CAPACITY NOTES:_____
_____

HIVE INSPECTION NOTES:

BEE MOOD/TEMPERAMENT:_____
POLLEN?_____   HONEY FLOW? _____
FOOD/WATER NOTES: _____
QUEEN? YES ☐ NO ☐   MARKED?: YES ☐ NO ☐   COLOR: _____
BROOD STAGE (EGG, LARVA, PUPA), PATTERN AND NOTES: _____
_____
_____

LARVAE STATUS AND NOTES: _____
_____
_____

COMB ABNORMALITIES (QUEEN CELLS/DRONE COMB BUILDING, ETC.):
_____
_____

SIGNS OF PESTS/PROBLEMS? (MITES, ANTS, MOTHS, DEAD BEES, SMELL):
_____
_____

TREATMENTS/MEDICATIONS: _____
_____

HONEY STORES: _____
SEASONAL NOTES (BLOOM, POLLEN/NECTAR SOURCES):_____
_____

RATING OF OVERALL HIVE HEALTH   ⬡ ⬡ ⬡ ⬡ ⬡
                                    WEAK   AVERAGE   STRONG

NOTES/COMB DIAGRAM:

BEEKEEPING JOURNAL/LOG   HIVE ID: _____

DATE: _____   TIME: _____   WEATHER: _____

HIVE STRUCTURE: FRAMES: _____   SUPERS: _____

BROOD FRAMES: _____   HONEY FRAMES: _____   OPEN FRAMES: _____

HIVE CAPACITY NOTES: _____
_____

HIVE INSPECTION NOTES:

BEE MOOD/TEMPERAMENT: _____
POLLEN? _____   HONEY FLOW? _____
FOOD/WATER NOTES: _____
QUEEN? YES ☐ NO ☐   MARKED?: YES ☐ NO ☐   COLOR: _____
BROOD STAGE (EGG, LARVA, PUPA), PATTERN AND NOTES: _____
_____
_____

LARVAE STATUS AND NOTES: _____
_____
_____

COMB ABNORMALITIES (QUEEN CELLS/DRONE COMB BUILDING, ETC.):
_____
_____

SIGNS OF PESTS/PROBLEMS? (MITES, ANTS, MOTHS, DEAD BEES, SMELL):
_____
_____

TREATMENTS/MEDICATIONS: _____
_____
_____

HONEY STORES: _____
SEASONAL NOTES (BLOOM, POLLEN/NECTAR SOURCES): _____
_____

RATING OF OVERALL HIVE HEALTH   ⬡ ⬡ ⬡ ⬡
                                   WEAK   AVERAGE   STRONG

NOTES/COMB DIAGRAM:

BEEKEEPING JOURNAL/LOG   HIVE ID: _____

DATE: _____   TIME: _____   WEATHER: _____

HIVE STRUCTURE: FRAMES: _____   SUPERS: _____

BROOD FRAMES: _____   HONEY FRAMES: _____   OPEN FRAMES: _____

HIVE CAPACITY NOTES: _____
_____

HIVE INSPECTION NOTES:

BEE MOOD/TEMPERAMENT: _____
POLLEN? _____   HONEY FLOW? _____
FOOD/WATER NOTES: _____
QUEEN? YES ☐ NO ☐   MARKED?: YES ☐ NO ☐   COLOR: _____
BROOD STAGE (EGG, LARVA, PUPA), PATTERN AND NOTES: _____
_____
_____

LARVAE STATUS AND NOTES: _____
_____
_____

COMB ABNORMALITIES (QUEEN CELLS/DRONE COMB BUILDING, ETC.):
_____
_____

SIGNS OF PESTS/PROBLEMS? (MITES, ANTS, MOTHS, DEAD BEES, SMELL):
_____
_____

TREATMENTS/MEDICATIONS: _____
_____
_____

HONEY STORES: _____
SEASONAL NOTES (BLOOM, POLLEN/NECTAR SOURCES): _____
_____

RATING OF OVERALL HIVE HEALTH   ⬡   ⬡   ⬡   ⬡   ⬡
                                   WEAK   AVERAGE   STRONG

NOTES/COMB DIAGRAM:

BEEKEEPING JOURNAL/LOG   HIVE ID: _____

DATE: _____   TIME: _____   WEATHER: _____

HIVE STRUCTURE: FRAMES: _____   SUPERS: _____

BROOD FRAMES: _____   HONEY FRAMES: _____   OPEN FRAMES: _____

HIVE CAPACITY NOTES: _____

_____

HIVE INSPECTION NOTES:

BEE MOOD/TEMPERAMENT: _____

POLLEN: _____   HONEY FLOW? _____

FOOD/WATER NOTES: _____

QUEEN? YES ☐ NO ☐   MARKED?: YES ☐ NO ☐   COLOR: _____

BROOD STAGE (EGG, LARVA, PUPA), PATTERN AND NOTES: _____

_____

_____

LARVAE STATUS AND NOTES: _____

_____

_____

COMB ABNORMALITIES (QUEEN CELLS/DRONE COMB BUILDING, ETC.):

_____

_____

SIGNS OF PESTS/PROBLEMS? (MITES, ANTS, MOTHS, DEAD BEES, SMELL):

_____

_____

TREATMENTS/MEDICATIONS: _____

_____

_____

HONEY STORES: _____

SEASONAL NOTES (BLOOM, POLLEN/NECTAR SOURCES): _____

_____

RATING OF OVERALL HIVE HEALTH   ⬡   ⬡   ⬡   ⬡   ⬡

WEAK   AVERAGE   STRONG

NOTES/COMB DIAGRAM:

BEEKEEPING JOURNAL/LOG   HIVE ID: _____

DATE: _____   TIME: _____   WEATHER: _____

HIVE STRUCTURE: FRAMES: _____   SUPERS: _____

BROOD FRAMES: _____   HONEY FRAMES: _____   OPEN FRAMES: _____

HIVE CAPACITY NOTES: _____
_____

HIVE INSPECTION NOTES:

BEE MOOD/TEMPERAMENT: _____
POLLEN? _____   HONEY FLOW? _____
FOOD/WATER NOTES: _____
QUEEN? YES ☐ NO ☐   MARKED?: YES ☐ NO ☐   COLOR: _____
BROOD STAGE (EGG, LARVA, PUPA), PATTERN AND NOTES: _____
_____
_____

LARVAE STATUS AND NOTES: _____
_____
_____

COMB ABNORMALITIES (QUEEN CELLS/DRONE COMB BUILDING, ETC.):
_____
_____

SIGNS OF PESTS/PROBLEMS? (MITES, ANTS, MOTHS, DEAD BEES, SMELL):
_____
_____

TREATMENTS/MEDICATIONS: _____
_____
_____

HONEY STORES: _____
SEASONAL NOTES (BLOOM, POLLEN/NECTAR SOURCES): _____
_____

RATING OF OVERALL HIVE HEALTH   ⬡   ⬡   ⬡   ⬡
                                       WEAK   AVERAGE   STRONG

NOTES/COMB DIAGRAM:

BEEKEEPING JOURNAL/LOG   HIVE ID:_____

DATE: _____ TIME: _____ WEATHER: _____

HIVE STRUCTURE: FRAMES:_____ SUPERS: _____

BROOD FRAMES: _____ HONEY FRAMES: _____ OPEN FRAMES: _____

HIVE CAPACITY NOTES:_____
_____

HIVE INSPECTION NOTES:

BEE MOOD/TEMPERAMENT: _____
POLLEN?_____ HONEY FLOW? _____
FOOD/WATER NOTES: _____
QUEEN? YES ☐ NO ☐   MARKED?: YES ☐ NO ☐   COLOR: _____
BROOD STAGE (EGG, LARVA, PUPA), PATTERN AND NOTES: _____
_____
_____

LARVAE STATUS AND NOTES: _____
_____
_____

COMB ABNORMALITIES (QUEEN CELLS/DRONE COMB BUILDING, ETC.):
_____
_____

SIGNS OF PESTS/PROBLEMS? (MITES, ANTS, MOTHS, DEAD BEES, SMELL):
_____
_____

TREATMENTS/MEDICATIONS: _____
_____

HONEY STORES: _____
SEASONAL NOTES (BLOOM, POLLEN/NECTAR SOURCES):_____
_____

RATING OF OVERALL HIVE HEALTH   ☐   ☐   ☐   ☐   ☐
                                       WEAK   AVERAGE   STRONG

NOTES/COMB DIAGRAM:

BEEKEEPING JOURNAL/LOG   HIVE ID: _____

DATE: _____   TIME: _____   WEATHER: _____

HIVE STRUCTURE: FRAMES: _____   SUPERS: _____

BROOD FRAMES: _____   HONEY FRAMES: _____   OPEN FRAMES: _____

HIVE CAPACITY NOTES: _____

_____

HIVE INSPECTION NOTES:

BEE MOOD/TEMPERAMENT: _____

POLLEN? _____   HONEY FLOW? _____

FOOD/WATER NOTES: _____

QUEEN? YES ☐ NO ☐   MARKED?: YES ☐ NO ☐   COLOR: _____

BROOD STAGE (EGG, LARVA, PUPA), PATTERN AND NOTES: _____

_____

_____

LARVAE STATUS AND NOTES: _____

_____

_____

COMB ABNORMALITIES (QUEEN CELLS/DRONE COMB BUILDING, ETC.):

_____

_____

SIGNS OF PESTS/PROBLEMS? (MITES, ANTS, MOTHS, DEAD BEES, SMELL):

_____

_____

TREATMENTS/MEDICATIONS: _____

_____

_____

HONEY STORES: _____

SEASONAL NOTES (BLOOM, POLLEN/NECTAR SOURCES): _____

_____

RATING OF OVERALL HIVE HEALTH   ⬡ ⬡ ⬡ ⬡ ⬡

WEAK   AVERAGE   STRONG

NOTES/COMB DIAGRAM:

BEEKEEPING JOURNAL/LOG   HIVE ID: _____

DATE: _____ TIME: _____ WEATHER: _____

HIVE STRUCTURE: FRAMES: _____ SUPERS: _____

BROOD FRAMES: _____ HONEY FRAMES: _____ OPEN FRAMES: _____

HIVE CAPACITY NOTES: _____
_____

HIVE INSPECTION NOTES:

BEE MOOD/TEMPERAMENT: _____
POLLEN?_____ HONEY FLOW? _____
FOOD/WATER NOTES: _____
QUEEN? YES☐ NO☐   MARKED?: YES☐ NO☐   COLOR: _____
BROOD STAGE (EGG, LARVA, PUPA), PATTERN AND NOTES: _____
_____
_____

LARVAE STATUS AND NOTES: _____
_____
_____

COMB ABNORMALITIES (QUEEN CELLS/DRONE COMB BUILDING, ETC.):
_____
_____

SIGNS OF PESTS/PROBLEMS? (MITES, ANTS, MOTHS, DEAD BEES, SMELL):
_____
_____

TREATMENTS/MEDICATIONS: _____
_____
_____

HONEY STORES: _____
SEASONAL NOTES (BLOOM, POLLEN/NECTAR SOURCES): _____
_____

RATING OF OVERALL HIVE HEALTH   ⬡ ⬡ ⬡ ⬡ ⬡
                                  WEAK   AVERAGE   STRONG

NOTES/COMB DIAGRAM:

BEEKEEPING JOURNAL/LOG   HIVE ID: _____

DATE: _____   TIME: _____   WEATHER: _____

HIVE STRUCTURE: FRAMES: _____   SUPERS: _____

BROOD FRAMES: _____   HONEY FRAMES: _____   OPEN FRAMES: _____

HIVE CAPACITY NOTES: _____
_____

HIVE INSPECTION NOTES:

BEE MOOD/TEMPERAMENT: _____
POLLEN? _____   HONEY FLOW? _____
FOOD/WATER NOTES: _____
QUEEN? YES ☐ NO ☐   MARKED?: YES ☐ NO ☐   COLOR: _____
BROOD STAGE (EGG, LARVA, PUPA), PATTERN AND NOTES: _____
_____
_____

LARVAE STATUS AND NOTES: _____
_____
_____

COMB ABNORMALITIES (QUEEN CELLS/DRONE COMB BUILDING, ETC.):
_____
_____

SIGNS OF PESTS/PROBLEMS? (MITES, ANTS, MOTHS, DEAD BEES, SMELL):
_____
_____

TREATMENTS/MEDICATIONS: _____
_____
_____

HONEY STORES: _____
SEASONAL NOTES (BLOOM, POLLEN/NECTAR SOURCES): _____
_____
_____

RATING OF OVERALL HIVE HEALTH   ⬡ ⬡ ⬡ ⬡ ⬡

WEAK   AVERAGE   STRONG

NOTES/COMB DIAGRAM:

BEEKEEPING JOURNAL/LOG   HIVE ID: _____

DATE: _____   TIME: _____   WEATHER: _____

HIVE STRUCTURE: FRAMES: _____   SUPERS: _____
BROOD FRAMES: _____   HONEY FRAMES: _____   OPEN FRAMES: _____
HIVE CAPACITY NOTES: _____
_____

HIVE INSPECTION NOTES:
BEE MOOD/TEMPERAMENT: _____
POLLEN? _____   HONEY FLOW? _____
FOOD/WATER NOTES: _____
QUEEN? YES ☐ NO ☐   MARKED? YES ☐ NO ☐   COLOR: _____
BROOD STAGE (EGG, LARVA, PUPA), PATTERN AND NOTES: _____
_____
_____

LARVAE STATUS AND NOTES: _____
_____
_____

COMB ABNORMALITIES (QUEEN CELLS/DRONE COMB BUILDING, ETC.):
_____
_____

SIGNS OF PESTS/PROBLEMS? (MITES, ANTS, MOTHS, DEAD BEES, SMELL):
_____
_____

TREATMENTS/MEDICATIONS: _____
_____
_____

HONEY STORES: _____

SEASONAL NOTES (BLOOM, POLLEN/NECTAR SOURCES): _____
_____
_____

RATING OF OVERALL HIVE HEALTH   ⬡   ⬡   ⬡   ⬡   ⬡
                                        WEAK   AVERAGE   STRONG

NOTES/COMB DIAGRAM:

BEEKEEPING JOURNAL/LOG   HIVE ID: _____

DATE: _____   TIME: _____   WEATHER: _____

HIVE STRUCTURE: FRAMES: _____   SUPERS: _____
BROOD FRAMES: _____   HONEY FRAMES: _____   OPEN FRAMES: _____
HIVE CAPACITY NOTES: _____
_____

HIVE INSPECTION NOTES:
BEE MOOD/TEMPERAMENT: _____
POLLEN? _____   HONEY FLOW? _____
FOOD/WATER NOTES: _____
QUEEN? YES ☐ NO ☐   MARKED? YES ☐ NO ☐   COLOR: _____
BROOD STAGE (EGG, LARVA, PUPA), PATTERN AND NOTES: _____
_____
_____

LARVAE STATUS AND NOTES: _____
_____
_____

COMB ABNORMALITIES (QUEEN CELLS/DRONE COMB BUILDING, ETC.):
_____
_____

SIGNS OF PESTS/PROBLEMS? (MITES, ANTS, MOTHS, DEAD BEES, SMELL):
_____
_____

TREATMENTS/MEDICATIONS: _____
_____
_____

HONEY STORES: _____
SEASONAL NOTES (BLOOM, POLLEN/NECTAR SOURCES): _____
_____

RATING OF OVERALL HIVE HEALTH   ⬡ ⬡ ⬡ ⬡ ⬡
WEAK   AVERAGE   STRONG

NOTES/COMB DIAGRAM:

BEEKEEPING JOURNAL/LOG   HIVE ID: _____

DATE: _____   TIME: _____   WEATHER: _____

HIVE STRUCTURE: FRAMES: _____   SUPERS: _____

BROOD FRAMES: _____   HONEY FRAMES: _____   OPEN FRAMES: _____

HIVE CAPACITY NOTES: _____

_____

HIVE INSPECTION NOTES:

BEE MOOD/TEMPERAMENT: _____

POLLEN? _____   HONEY FLOW? _____

FOOD/WATER NOTES: _____

QUEEN? YES ☐ NO ☐   MARKED?: YES ☐ NO ☐   COLOR: _____

BROOD STAGE (EGG, LARVA, PUPA), PATTERN AND NOTES: _____

_____

_____

LARVAE STATUS AND NOTES: _____

_____

_____

COMB ABNORMALITIES (QUEEN CELLS/DRONE COMB BUILDING, ETC.): _____

_____

_____

SIGNS OF PESTS/PROBLEMS? (MITES, ANTS, MOTHS, DEAD BEES, SMELL): ___

_____

_____

TREATMENTS/MEDICATIONS: _____

_____

_____

HONEY STORES: _____

SEASONAL NOTES (BLOOM, POLLEN/NECTAR SOURCES): _____

_____

_____

RATING OF OVERALL HIVE HEALTH   ◯ ◯ ◯ ◯ ◯
                                   WEAK   AVERAGE   STRONG

NOTES/COMB DIAGRAM:

BEEKEEPING JOURNAL/LOG   HIVE ID: _____

DATE: _____   TIME: _____   WEATHER: _____

HIVE STRUCTURE: FRAMES: _____   SUPERS: _____
BROOD FRAMES: _____   HONEY FRAMES: _____   OPEN FRAMES: _____
HIVE CAPACITY NOTES: _____
_____

HIVE INSPECTION NOTES:
BEE MOOD/TEMPERAMENT: _____
POLLEN? _____   HONEY FLOW? _____
FOOD/WATER NOTES: _____
QUEEN? YES ☐ NO ☐   MARKED?: YES ☐ NO ☐   COLOR: _____
BROOD STAGE (EGG, LARVA, PUPA), PATTERN AND NOTES: _____
_____
_____

LARVAE STATUS AND NOTES: _____
_____
_____

COMB ABNORMALITIES (QUEEN CELLS/DRONE COMB BUILDING, ETC.):
_____
_____

SIGNS OF PESTS/PROBLEMS? (MITES, ANTS, MOTHS, DEAD BEES, SMELL):
_____
_____

TREATMENTS/MEDICATIONS: _____
_____
_____

HONEY STORES: _____
SEASONAL NOTES (BLOOM, POLLEN/NECTAR SOURCES): _____
_____

RATING OF OVERALL HIVE HEALTH   ⬡ ⬡ ⬡ ⬡ ⬡
WEAK   AVERAGE   STRONG

NOTES/COMB DIAGRAM:

BEEKEEPING JOURNAL/LOG   HIVE ID: _____

DATE: _____   TIME: _____   WEATHER: _____

HIVE STRUCTURE: FRAMES: _____   SUPERS: _____

BROOD FRAMES: _____   HONEY FRAMES: _____   OPEN FRAMES: _____

HIVE CAPACITY NOTES: _____
_____

HIVE INSPECTION NOTES:

BEE MOOD/TEMPERAMENT: _____
POLLEN? _____   HONEY FLOW? _____
FOOD/WATER NOTES: _____
QUEEN? YES ☐ NO ☐   MARKED?: YES ☐ NO ☐   COLOR: _____
BROOD STAGE (EGG, LARVA, PUPA), PATTERN AND NOTES: _____
_____
_____

LARVAE STATUS AND NOTES: _____
_____
_____

COMB ABNORMALITIES (QUEEN CELLS/DRONE COMB BUILDING, ETC.):
_____
_____

SIGNS OF PESTS/PROBLEMS? (MITES, ANTS, MOTHS, DEAD BEES, SMELL):
_____
_____

TREATMENTS/MEDICATIONS: _____
_____
_____

HONEY STORES: _____
SEASONAL NOTES (BLOOM, POLLEN/NECTAR SOURCES): _____
_____

RATING OF OVERALL HIVE HEALTH   ⬡ ⬡ ⬡ ⬡ ⬡
                                   WEAK   AVERAGE   STRONG

NOTES/COMB DIAGRAM:

BEEKEEPING JOURNAL/LOG HIVE ID: _____

DATE: _____ TIME: _____ WEATHER: _____

HIVE STRUCTURE: FRAMES: _____ SUPERS: _____

BROOD FRAMES: _____ HONEY FRAMES: _____ OPEN FRAMES: _____

HIVE CAPACITY NOTES: _____
_____

HIVE INSPECTION NOTES:

BEE MOOD/TEMPERAMENT: _____
POLLEN? _____ HONEY FLOW? _____
FOOD/WATER NOTES: _____
QUEEN? YES ☐ NO ☐   MARKED?: YES ☐ NO ☐   COLOR: _____
BROOD STAGE (EGG, LARVA, PUPA), PATTERN AND NOTES: _____
_____
_____

LARVAE STATUS AND NOTES: _____
_____
_____

COMB ABNORMALITIES (QUEEN CELLS/DRONE COMB BUILDING, ETC.):
_____
_____

SIGNS OF PESTS/PROBLEMS? (MITES, ANTS, MOTHS, DEAD BEES, SMELL):
_____
_____

TREATMENTS/MEDICATIONS: _____
_____
_____

HONEY STORES: _____
SEASONAL NOTES (BLOOM, POLLEN/NECTAR SOURCES): _____
_____

RATING OF OVERALL HIVE HEALTH   ⬡   ⬡   ⬡   ⬡   ⬡
                                  WEAK   AVERAGE   STRONG

NOTES/COMB DIAGRAM:

BEEKEEPING JOURNAL/LOG HIVE ID: _____

DATE: _____ TIME: _____ WEATHER: _____

HIVE STRUCTURE: FRAMES: _____ SUPERS: _____

BROOD FRAMES: _____ HONEY FRAMES: _____ OPEN FRAMES: _____

HIVE CAPACITY NOTES: _____
_____

HIVE INSPECTION NOTES:

BEE MOOD/TEMPERAMENT: _____
POLLEN? _____ HONEY FLOW? _____
FOOD/WATER NOTES: _____
QUEEN? YES ☐ NO ☐   MARKED?: YES ☐ NO ☐   COLOR: _____
BROOD STAGE (EGG, LARVA, PUPA), PATTERN AND NOTES: _____
_____
_____

LARVAE STATUS AND NOTES: _____
_____
_____

COMB ABNORMALITIES (QUEEN CELLS/DRONE COMB BUILDING, ETC.): _____
_____
_____

SIGNS OF PESTS/PROBLEMS? (MITES, ANTS, MOTHS, DEAD BEES, SMELL): ___
_____
_____

TREATMENTS/MEDICATIONS: _____
_____

HONEY STORES: _____
SEASONAL NOTES (BLOOM, POLLEN/NECTAR SOURCES): _____
_____

RATING OF OVERALL HIVE HEALTH   ◯ ◯ ◯ ◯ ◯
                                  WEAK   AVERAGE   STRONG

NOTES/COMB DIAGRAM:

BEEKEEPING JOURNAL/LOG HIVE ID: _____

DATE: _____ TIME: _____ WEATHER: _____

HIVE STRUCTURE: FRAMES: _____ SUPERS: _____

BROOD FRAMES: _____ HONEY FRAMES: _____ OPEN FRAMES: _____

HIVE CAPACITY NOTES: _____

_____

HIVE INSPECTION NOTES:

BEE MOOD/TEMPERAMENT: _____

POLLEN? _____ HONEY FLOW? _____

FOOD/WATER NOTES: _____

QUEEN? YES ☐ NO ☐   MARKED?: YES ☐ NO ☐   COLOR: _____

BROOD STAGE (EGG, LARVA, PUPA), PATTERN AND NOTES: _____

_____

_____

LARVAE STATUS AND NOTES: _____

_____

_____

COMB ABNORMALITIES (QUEEN CELLS/DRONE COMB BUILDING, ETC.):

_____

_____

SIGNS OF PESTS/PROBLEMS? (MITES, ANTS, MOTHS, DEAD BEES, SMELL):

_____

_____

TREATMENTS/MEDICATIONS: _____

_____

_____

HONEY STORES: _____

SEASONAL NOTES (BLOOM, POLLEN/NECTAR SOURCES): _____

_____

RATING OF OVERALL HIVE HEALTH   ⬡ ⬡ ⬡ ⬡ ⬡
                                   WEAK   AVERAGE   STRONG

NOTES/COMB DIAGRAM:

BEEKEEPING JOURNAL/LOG   HIVE ID: _____

DATE: _____   TIME: _____   WEATHER: _____

HIVE STRUCTURE: FRAMES: _____   SUPERS: _____
BROOD FRAMES: _____   HONEY FRAMES: _____   OPEN FRAMES: _____
HIVE CAPACITY NOTES: _____
_____

HIVE INSPECTION NOTES:
BEE MOOD/TEMPERAMENT: _____
POLLEN? _____   HONEY FLOW? _____
FOOD/WATER NOTES: _____
QUEEN? YES ☐ NO ☐   MARKED?: YES ☐ NO ☐   COLOR: _____
BROOD STAGE (EGG, LARVA, PUPA), PATTERN AND NOTES: _____
_____
_____

LARVAE STATUS AND NOTES: _____
_____
_____

COMB ABNORMALITIES (QUEEN CELLS/DRONE COMB BUILDING, ETC.):
_____
_____

SIGNS OF PESTS/PROBLEMS? (MITES, ANTS, MOTHS, DEAD BEES, SMELL):
_____
_____

TREATMENTS/MEDICATIONS: _____
_____
_____

HONEY STORES: _____
SEASONAL NOTES (BLOOM, POLLEN/NECTAR SOURCES): _____
_____
_____

RATING OF OVERALL HIVE HEALTH   ⬡ ⬡ ⬡ ⬡ ⬡
WEAK   AVERAGE   STRONG

NOTES/COMB DIAGRAM:

BEEKEEPING JOURNAL/LOG   HIVE ID: _____

DATE: _____ TIME: _____ WEATHER: _____

HIVE STRUCTURE: FRAMES: _____ SUPERS: _____

BROOD FRAMES: _____ HONEY FRAMES: _____ OPEN FRAMES: _____

HIVE CAPACITY NOTES: _____
_____

HIVE INSPECTION NOTES:

BEE MOOD/TEMPERAMENT: _____

POLLEN? _____ HONEY FLOW? _____

FOOD/WATER NOTES: _____

QUEEN? YES ☐ NO ☐   MARKED?: YES ☐ NO ☐   COLOR: _____

BROOD STAGE (EGG, LARVA, PUPA), PATTERN AND NOTES: _____
_____
_____

LARVAE STATUS AND NOTES: _____
_____
_____

COMB ABNORMALITIES (QUEEN CELLS/DRONE COMB BUILDING, ETC.):
_____
_____

SIGNS OF PESTS/PROBLEMS? (MITES, ANTS, MOTHS, DEAD BEES, SMELL):
_____
_____

TREATMENTS/MEDICATIONS: _____
_____

HONEY STORES: _____

SEASONAL NOTES (BLOOM, POLLEN/NECTAR SOURCES): _____
_____
_____

RATING OF OVERALL HIVE HEALTH   ⬡ ⬡ ⬡ ⬡ ⬡
                                  WEAK   AVERAGE   STRONG

BEEKEEPING INCOME

YEAR:_____ BALANCE BROUGHT FORWARD:_____

DATE	BUYER	DESCRIPTION	UNITS	COST/UNIT	TOTAL

BEEKEEPING INCOME

YEAR:_____ BALANCE BROUGHT FORWARD:_____

DATE	BUYER	DESCRIPTION	UNITS	COST/UNIT	TOTAL

BEEKEEPING EXPENSES

DATE	VENDOR	DESCRIPTION	UNITS	PRICE/UNIT	TOTAL

BEEKEEPING EXPENSES

DATE	VENDOR	DESCRIPTION	UNITS	PRICE/UNIT	TOTAL

EQUIPMENT INVENTORY

#	EQUIPMENT DESCRIPTION	SERIAL #	PURCHASE DATE	WARRANTY GOOD TILL	COST

EQUIPMENT USE/CLEANING

EQUIPMENT DESCRIPTION	DATE USED	DATE CLEANED	CLEANED BY	NOTES